# THE GREAT VOYAGERS

## Earth's Intergalactic Ambassadors

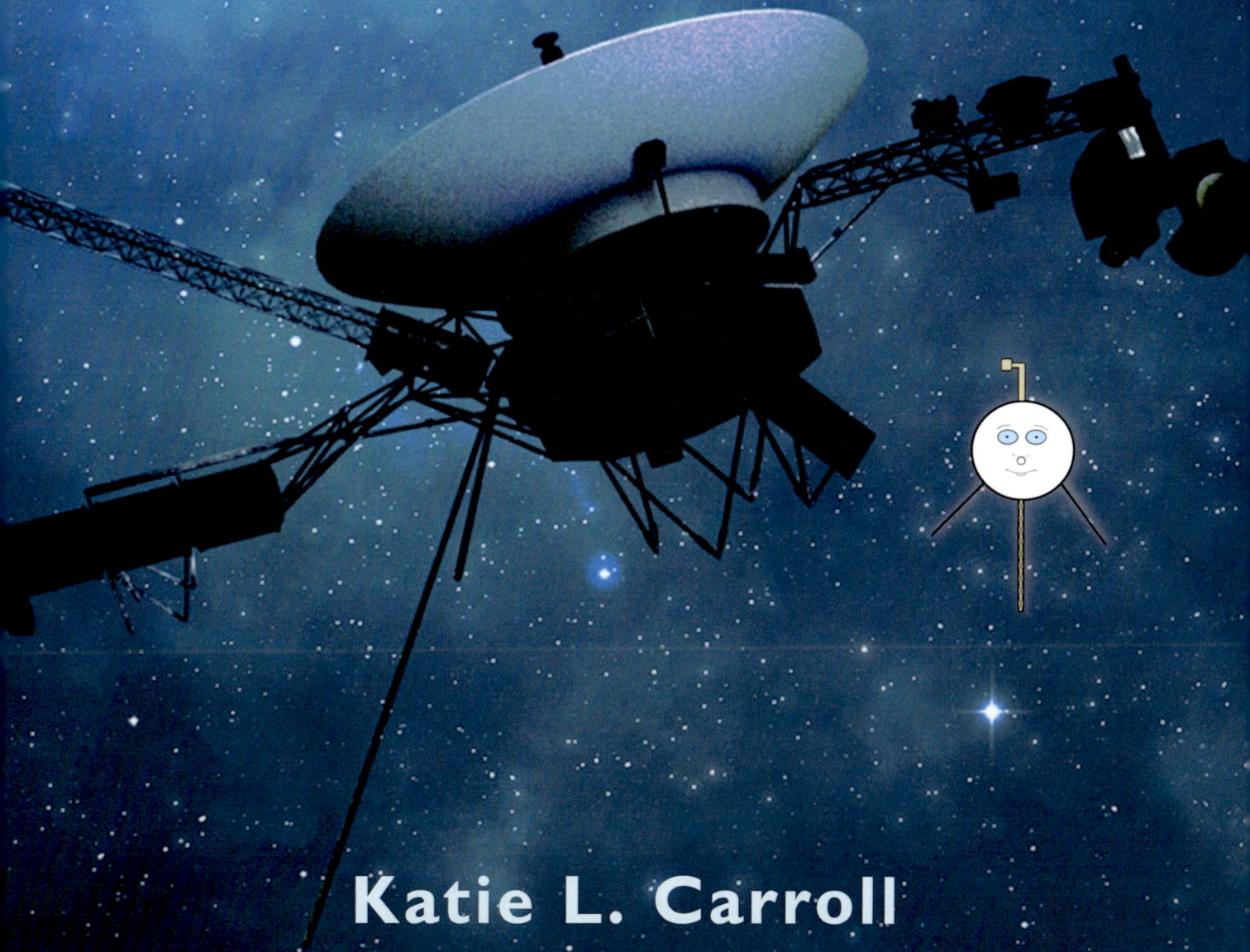

## Katie L. Carroll

A rare planetary alignment sparks a mission:

A grand tour of the outer planets.

Two bold Voyagers slingshot through the solar system,

Embarking on a journey of firsts.

A montage of Jupiter,
Saturn, Uranus, and
Neptune images taken
by Voyager 2.

The missions of NASA spacecraft Voyager 1 and Voyager 2 were made possible by two factors. First, the four outer planets—Jupiter, Saturn, Uranus, and Neptune—would be in alignment so that a craft launched in the late 1970s could visit them all. Second, scientists proved that a gravity assist maneuver was possible. A gravity assist maneuver uses a planet's gravity to slingshot a spacecraft to its next location, saving energy for longer space travel.

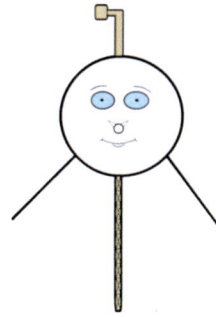

An artist's rendering of a Voyager spacecraft launching from Earth, passing Jupiter, and performing a gravity assist maneuver on its way to Saturn.

The Voyagers flew way out past the clouds,

Past the Moon, marked by the American flag and footprints,

Beyond Mars with its red, iron-oxide surface,

And through the asteroid belt.

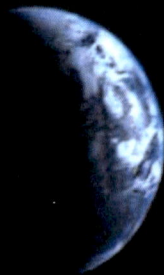

Voyager 1 was 7.25 million miles from Earth when it recorded this crescent Earth and Moon on September 18, 1977.

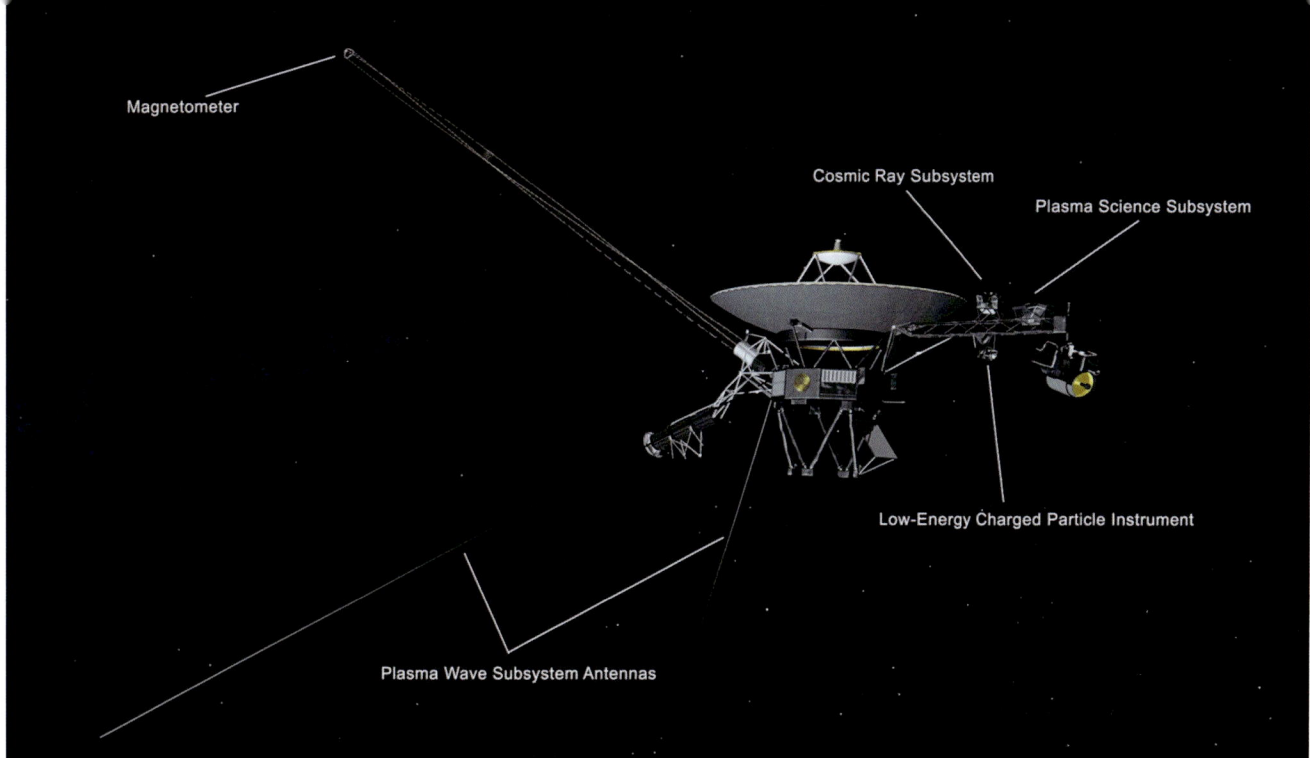

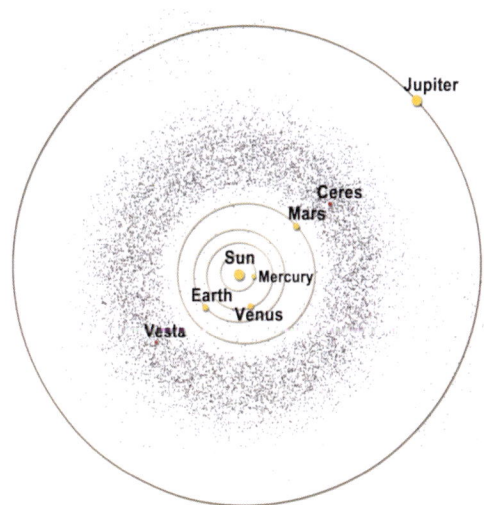

Top: An illustration of select science instruments on the Voyagers.
Bottom left: Astronaut Edwin E. Aldrin, Jr. on the moon.
Bottom right: An artist's rendering of the asteroid belt.

Journeying to the gas giants and their many moons,
Both Voyagers explored stormy Jupiter with volcanic Io and icy Europa.

Voyager 1 reached its closest point to Jupiter in March 1979, with Voyager 2 following four months later. The data and images revealed that Jupiter's Great Red Spot was similar to a very large hurricane. Lightning was detected on Jupiter, the first time it was seen anywhere other than Earth. Other firsts included discovering Jupiter's ring system, observing volcanoes on the moon Io, and finding cracked ice over a liquid ocean on the moon Europa.

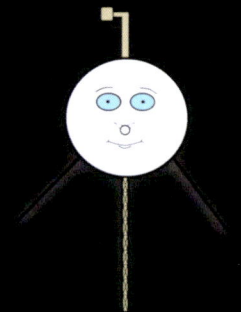

They analyzed ringed Saturn
and its atmospheric moon Titan.

Then the great Voyagers'
paths diverged.

In November 1980,
Voyager 1 approached Saturn
and the planet's largest moon, Titan,
while Voyager 2 reached Saturn in
August 1981. The Voyagers uncovered
new moons and a new ring. Scientists
theorized that the rings were formed
from shattered moons. The Voyagers
revealed that Titan's atmosphere
is rich with nitrogen, much like
Earth's atmosphere.

An image of Saturn with
moons Dione and Rhea,
taken by Voyager 2 on
July 21, 1981.

# Moons

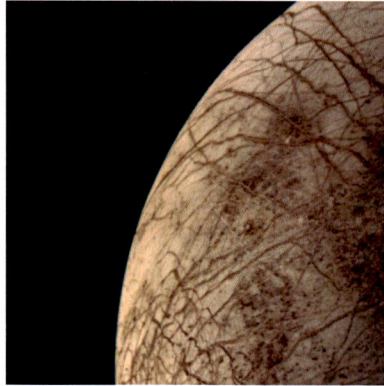

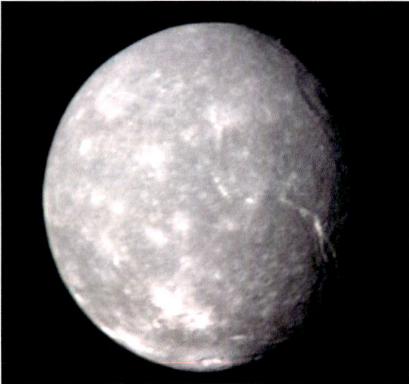

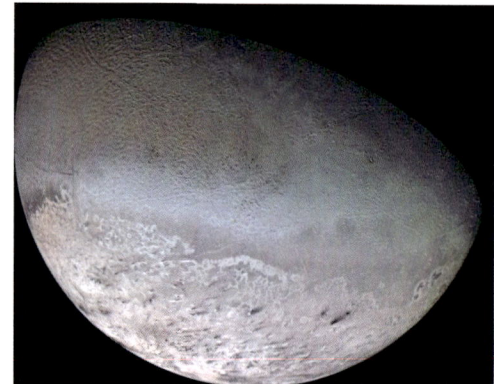

Though Voyager 1's planetary mission ended after Saturn, its overall mission was far from over. Meanwhile, Voyager 2 was approved to explore Uranus and Neptune. In early 1986, it made its closest approach to Uranus, the first time humans were able to observe this planet up close. Images and measurements revealed 11 new moons, the planet's tilted magnetic field, and the coldest temperatures in the solar system at minus 353 degrees F (213.9 degrees C).

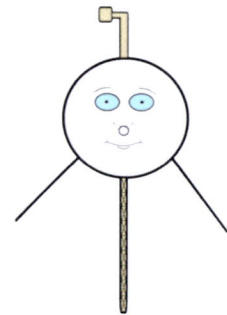

While Voyager 1 drifted deeper into space,

Voyager 2 continued to the distant ice giants.

A trip by tilted Uranus with its dark rings
and freezing temperatures,

And a flyby of windy Neptune with
The Great Dark Spot.

The final planetary encounter for Voyager 2 came in August 1989, nearly 12 years after lift-off. The Neptune data revealed six new moons and a huge storm called The Great Dark Spot. With these observations, Voyager 2 became the first spacecraft to visit all four outer planets. Shortly after, scientists turned off Voyager 2's cameras to conserve power for the other science instruments on board.

Left: An image of Uranus taken by Voyager 2 in 1986. Right: An image of Neptune showing The Great Dark Spot, created from images taken by Voyager 2 in 1989.

For more than 45 years, the Voyagers have persevered.

Many billions of miles from Earth,

They maintain communications and share new discoveries

With the tiny blue dot of their home planet.

VENUS

EARTH

JUPITER

SATURN

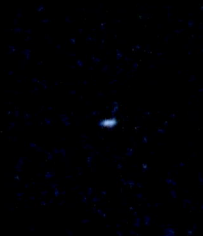
URANUS

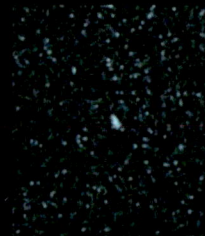
NEPTUNE

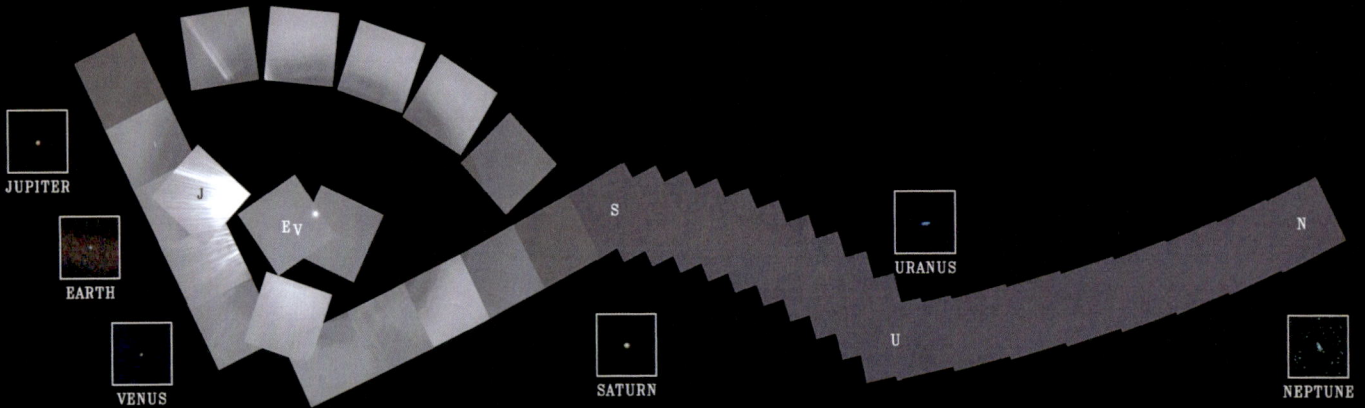

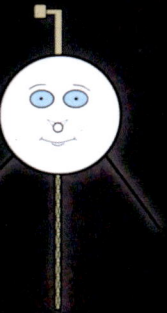

By February 1990, Voyager 1 was almost four billion miles away from the sun. Carl Sagan, renowned astronomer and a founding scientist of the Voyager missions, convinced the NASA team to point the cameras back toward Earth. In a series of images, Voyager 1 captured Venus, Earth, Jupiter, Saturn, Uranus, and Neptune in their positions around the sun. Mercury was too close to the sun, and Mars was too hard to see in the scattered sunlight. The collective image became known as the Solar System Family Portrait. Then, Voyager 1's cameras were turned off.

Earth as a very small dot lit by a sunbeam, prompting Carl Sagan to refer to it as the "pale blue dot."

These intrepid Voyagers navigated

Beyond the Kuiper Belt, home to the
dwarf planets,

Beyond the heliopause where solar winds
crash against interstellar space,

Beyond any other human-made objects.

An artist's rendering of
Voyager 1 entering the
orange glow of
interstellar space.

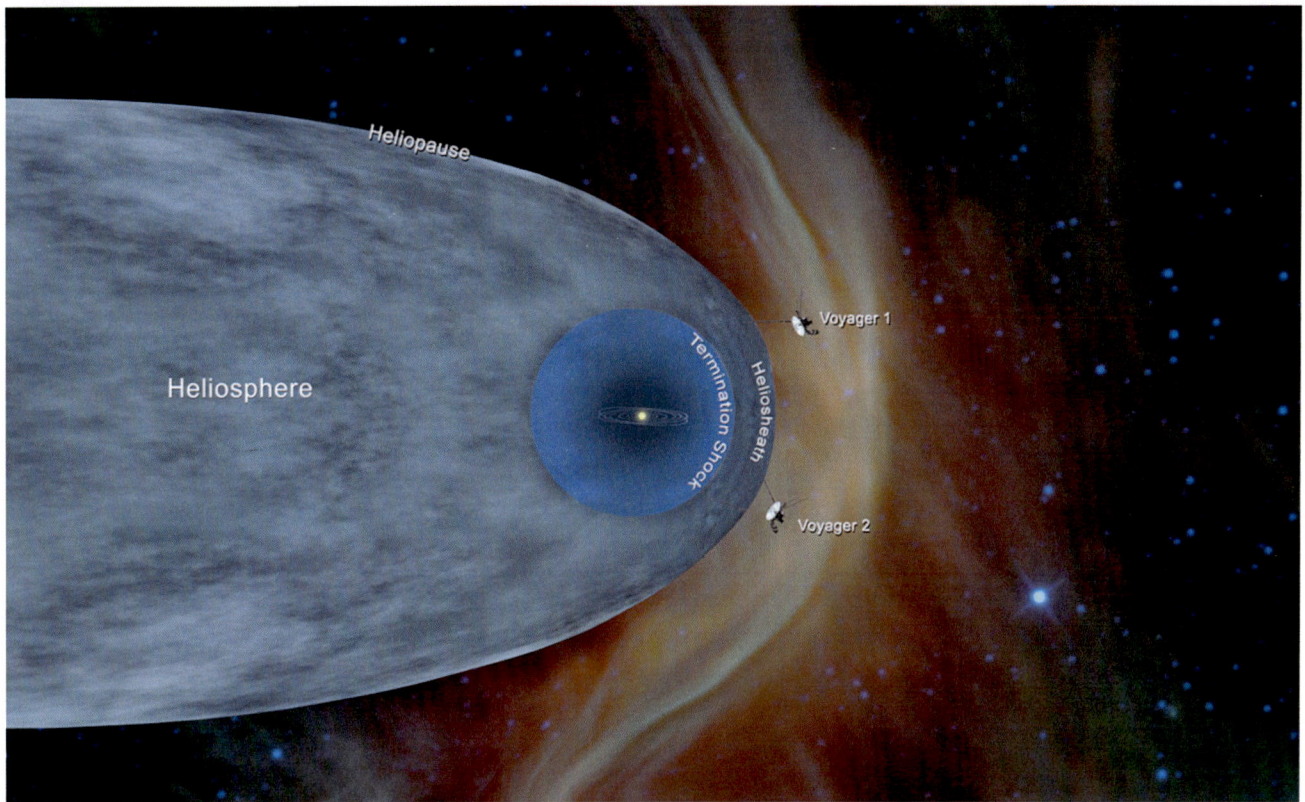

An illustration of the positions of Voyager 1 and Voyager 2 as they traveled beyond the heliopause and into interstellar space where solar winds meet the stellar winds of other stars.

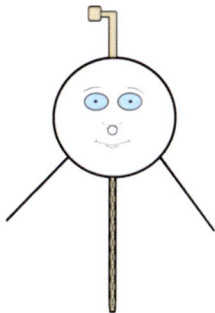

The Voyagers both reached milestones 35 years after lift-off. On August 13, 2012, Voyager 2 achieved 12,759 days of continuous operation and became the longest-running NASA mission. On August 25, 2012, Voyager 1 was the first human-made object to reach interstellar space when it passed the heliopause, the point where the sun's solar winds meet interstellar winds from other stars. It would take Voyager 2 another six years to cross the heliopause into interstellar space, achieving this on November 5, 2018.

But soon, they will have traveled too far to contact home.

Then they will make the long, dark journey through unchartered space alone,

Their next close encounters with the stars tens of thousands of years away,

Wandering the Milky Way for ages to come.

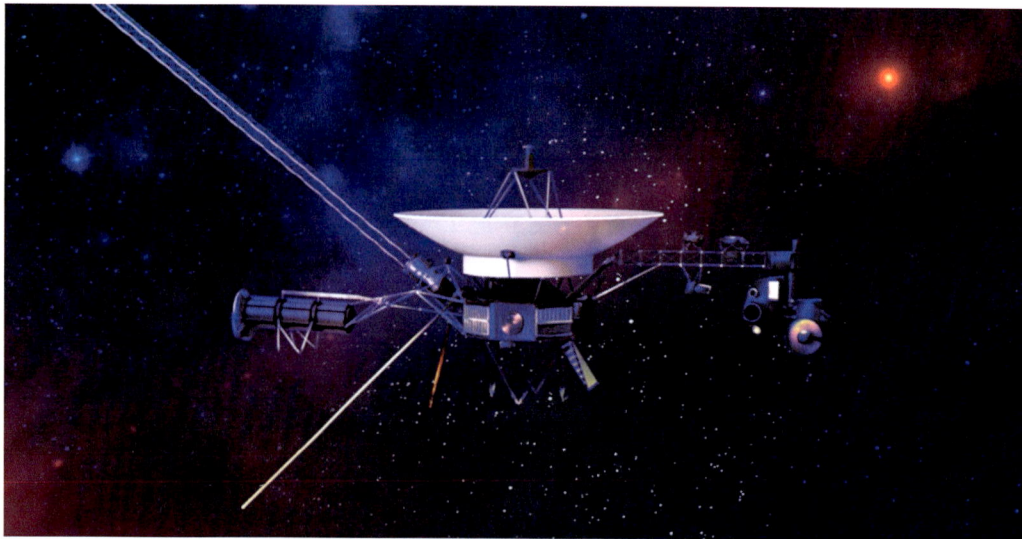

Above: An artist's rendering of the Voyager spacecraft.
Right: An artist's rendering of the Milky Way Galaxy.

Both Voyagers continue to travel through the Milky Way with Voyager 1 over 15 billion miles from Earth. With a limited number of scientific instruments operational, they continue to collect data beyond the heliopause. Power-saving measures are being taken to keep the active science mission going as long as possible. The Voyagers will be in communication range until about 2036.

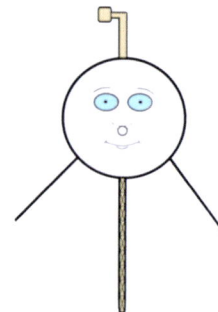

Even without human contact, their mission will go on.

A slice of humanity rides along in the form of

Two Golden Records, designed to last billions of years.

Records that contain pieces of what it's like to live on
the tiny blue dot of Earth.

A montage of the front and back of the Golden Record and a Voyager spacecraft.

About 40,000 years from now, Voyager 1 will pass within 1.6 light years of a star called Gliese 445 in the giraffe constellation Camelopardalis. Around the same time, Voyager 2 will be 1.7 light years from the star Ross 248. Both probes hold identical phonograph records that are gold-plated. On their covers, a symbolic language shows how they can be played. The contents are a limited record of human life on Earth for any extraterrestrials who might encounter the Voyagers.

An image of star Gliese 445, circled in red, taken by the Oschin Schmidt Telescope.

# Images of Earth

An astronaut on a spacewalk, a rocket launch, Earth from space;

Children touching a globe, a woman eating fruit;

A diagram of continental drift, DNA structure;

Dolphins, a river, a sunset with birds, snowflakes on a sequoia tree.

Top left: Astronaut Edward White on a spacewalk.
Bottom left: Earth from Apollo 11.
Right: First launch of the Titan Centaur.

# Languages of Earth

A woman in the Amoy dialect, "Friends of space, how are you all? Have you eaten yet? Come visit us if you have time."

A man in Spanish, "Hello and greetings to all."

A woman in Polish, "Welcome, creatures from beyond the outer world."

A child in English, "Hello from the children of planet Earth."

Tanti auguri e saluti
تحياتنا للأصدقاء في النجوم. يا ليت يجمعنا الزمان.
Salutări la toată lumea
안녕하세요

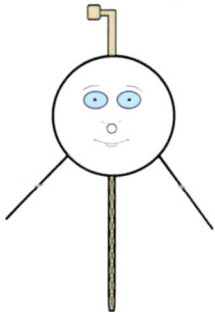

The Golden Records were created long before images could be stored in digital form, as is done on computers and cell phones today. For the 115 images, NASA invented a technology that converted the images into audio waves and embedded them onto the surface of the records. Under a tight deadline, the committee decided to include speech in 55 different languages. The speakers were simply instructed to record a brief greeting that one day might be heard by extraterrestrials.

# Sounds of Earth

Crickets chirping
and wild dogs howling,

Thunder rumbling
and waves crashing,

A train chugging
and a rocket blasting off,

Human laughter
and a mother soothing a crying baby.

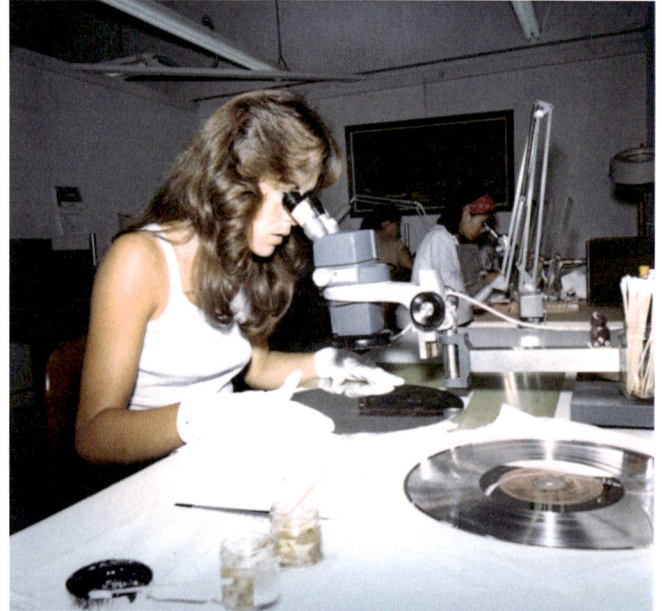

An inspection of the engraving of the mother record during the production process of the Golden Records, taken on July 28, 1977.

A series of sound recordings on the Golden Records includes humans, animals, natural phenomena, and human-made inventions. The last sound clip is a pulsar—radio waves from a rotating, densely compact star—as detected from Earth. The final section of audio is 90 minutes of music. The song clips represent classic and modern music (up to the late 1970s when the Voyagers launched), as well as music from all different parts of the world. Rather than a thorough tour of musical history, it's a sampling of what humans have to offer.

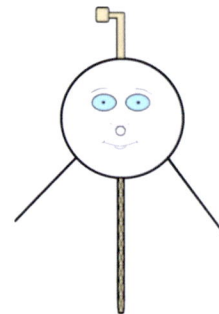

# Music of Earth

The classics of Beethoven's Fifth Symphony and Mozart's The Magic Flute,

The rock of Chuck Berry's "Johnny B. Goode,"

Pan pipes from the Soloman Islands,

A wedding song from Peru.

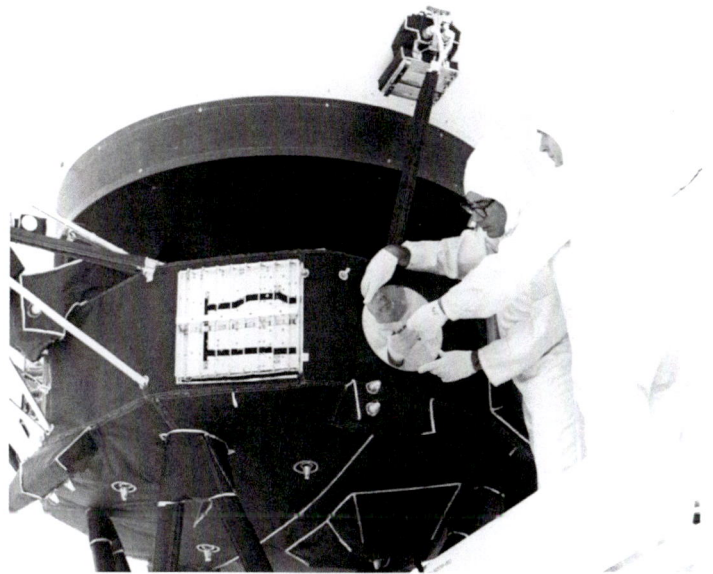

Left: Preparing the Voyager 1 Golden Record to be installed on the spacecraft. Right: Engineers secure the cover over the Golden Record on Voyager 1.

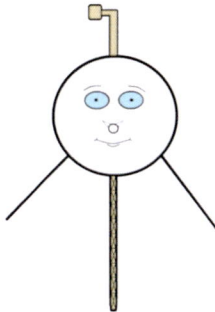

The Voyagers are Earth's first intergalactic ambassadors.

In the future, they may hold the last remaining evidence of human civilization,

A small piece of the life and history of the people here on Earth,

Perhaps for some galaxy-wandering beings to discover.

Some scientists believe the future recipients of the Golden Records will be descendants of humans, while others think they will never be found by intelligent life. Carl Sagan had this to say about the records, "The spacecraft will be encountered and the record played only if there are advanced spacefaring civilizations in interstellar space. But the launching of this bottle into the cosmic ocean says something very hopeful about life on this planet."

# Voyager Timeline

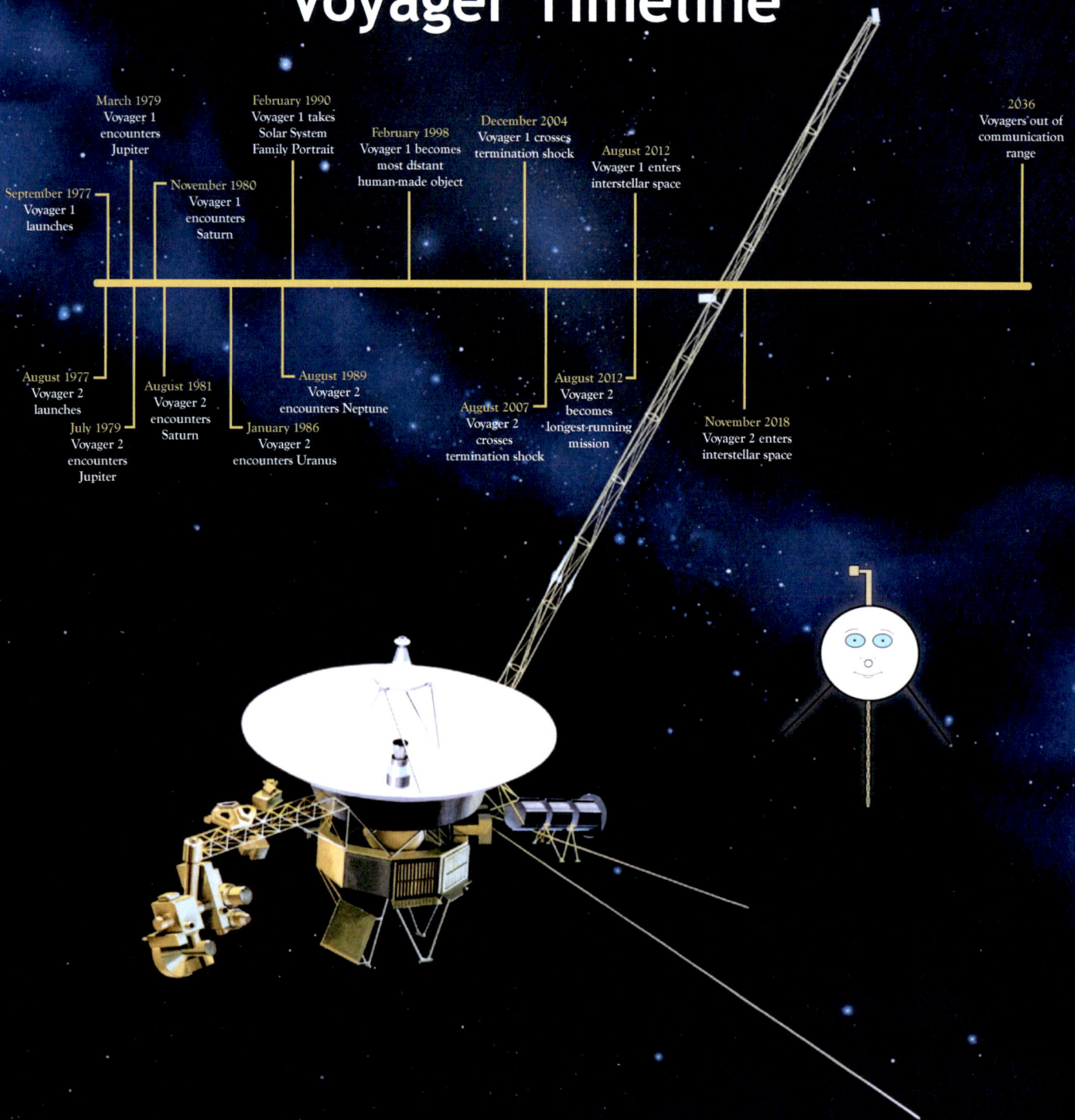

**March 1979**
Voyager 1 encounters Jupiter

**February 1990**
Voyager 1 takes Solar System Family Portrait

**February 1998**
Voyager 1 becomes most distant human-made object

**December 2004**
Voyager 1 crosses termination shock

**August 2012**
Voyager 1 enters interstellar space

**2036**
Voyagers out of communication range

**September 1977**
Voyager 1 launches

**November 1980**
Voyager 1 encounters Saturn

**August 1977**
Voyager 2 launches

**August 1981**
Voyager 2 encounters Saturn

**August 1989**
Voyager 2 encounters Neptune

**August 2007**
Voyager 2 crosses termination shock

**August 2012**
Voyager 2 becomes longest-running mission

**November 2018**
Voyager 2 enters interstellar space

**July 1979**
Voyager 2 encounters Jupiter

**January 1986**
Voyager 2 encounters Uranus

# Further Exploration

Voyager Mission Homepage
https://voyager.jpl.nasa.gov/

Golden Record
https://voyager.jpl.nasa.gov/golden-record/

Interactive Timeline
https://voyager.jpl.nasa.gov/mission/timeline/

Mission Animation
https://eyes.nasa.gov/apps/solar-system/#/story/voyager_grand_tour

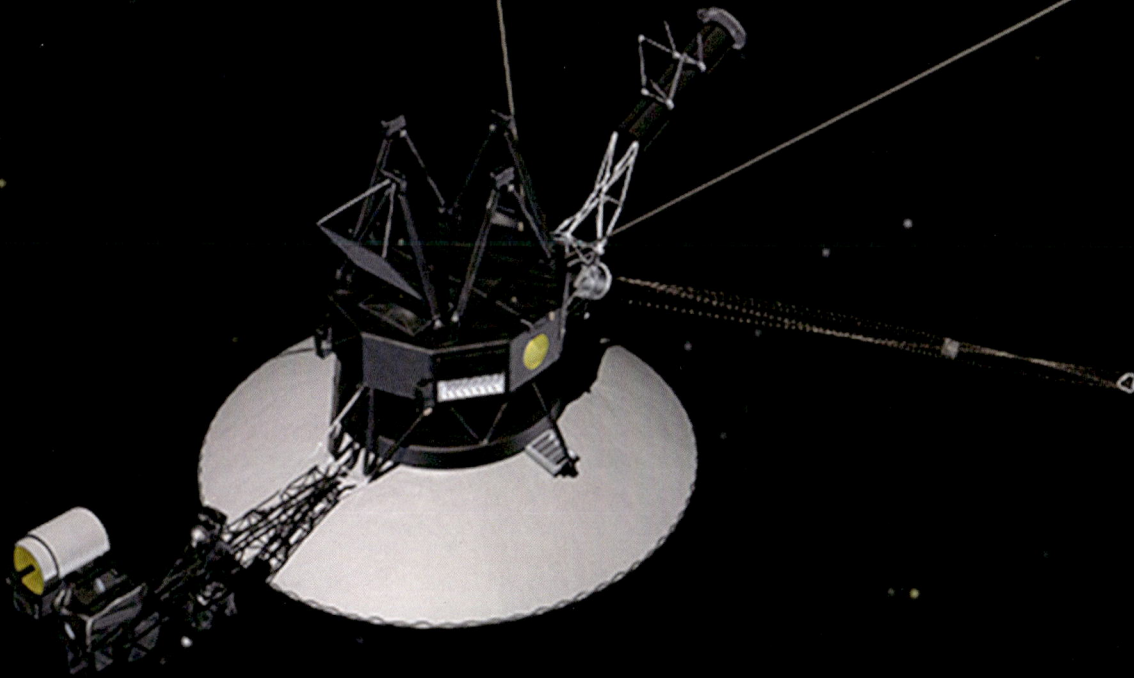

*Dedication: To those who dream to explore the far reaching places of the universe.*

## Photo Credits

All photos courtesy of NASA/JPL-Caltech except:
Caltech/Palomar: page 17
NASA: page 3, 5 (bottom left), 18 (top left, bottom left, right)
NASA/JPL: page 2, 4, 6, 7, 8 (top middle, top right, bottom left, bottom middle), 9 (right), 10 (top, bottom), 11, 15, 24 (background)
NASA/JPL/USGS: page 8 (top left, bottom right)
NASA/McREL: page 5 (bottom right)

Library of Congress Cataloguing-in-Publication Data
Name: Carroll, Katie L., author.
Title: The great voyagers: Earth's intergalactic ambassadors / Katie L. Carroll.
Identifiers: LCCN 2024914194 (print)   ISBN 9781958575116 (print)   ISBN 9781958575123 (ebook)
Subjects: 1. Space probes—Juvenile literature. 2. Outer space—Exploration—Juvenile literature.
LC record available at https://lccn.loc.gov/2024914194

Published by Shimmer Publications, LLC
Milford, Connecticut
Visit the author's website at www.katielcarroll.com

www.ingramcontent.com/pod-product-compliance
Lightning Source LLC
Chambersburg PA
CBRC090247230326
41458CB00108B/6514